Simple Machines

Wheels
and Axles

by Martha E. H. Rustad

CAPSTONE PRESS
a capstone imprint

Little Pebble is published by Capstone Press,
1710 Roe Crest Drive, North Mankato, Minnesota 56003
www.mycapstone.com

Library of Congress Cataloging-in-Publication Data
Names: Rustad, Martha E. H. (Martha Elizabeth Hillman), 1975– author.
Title: Wheels and axles / by Martha E.H. Rustad.
Description: North Mankato, Minnesota : Little Pebble is published by
    Capstone Press, [2018] | Series: Simple machines | Audience: Age 4–7.
Identifiers: LCCN 2017031571 (print) | LCCN 2017035876 (ebook) |
    ISBN 9781543500882 (eBook PDF) | ISBN 9781543500769 (hardcover) |
    ISBN 9781543500820 (paperback)
Subjects: LCSH: Wheels—Juvenile literature. | Axles—Juvenile literature.
Classification: LCC TJ181.5 (ebook) | LCC TJ181.5 .R88 2018 (print) | DDC
    621.8—dc23
LC record available at https://lccn.loc.gov/2017031571

Editorial Credits
Marissa Kirkman, editor; Kyle Grentz (cover) and Charmaine Whitman (interior), designers; Jo Miller, media researcher; Katy LaVigne, production specialist

Image Credits
Capstone Studio: Karon Dubke, 11; Dreamstime: Eti Swimford, cover, 1; Newscom: Stephen Welstead Blend Images/LWA, 19; Shutterstock: edwardolive, 9, Jan Toula, 22, maxbelchenko, 8, Mike Flippo, 7 (top), Monkey Business Images, 21, Secheltgirl, 5, smereka, 13, Vitalliy, 15, Vitpho, 7 (bottom), Waraporn Chokchaiworarat, 17

Design Elements
Capstone

Printed and bound in the USA.
010766S18

Table of Contents

Help with Work

Work is hard!

We need help.

Use a simple machine.

These tools help us work.

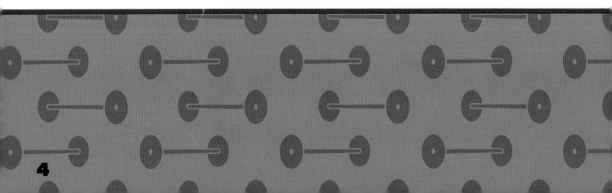

wheels and axles

Wheels and axles are
all around.
They move heavy loads.

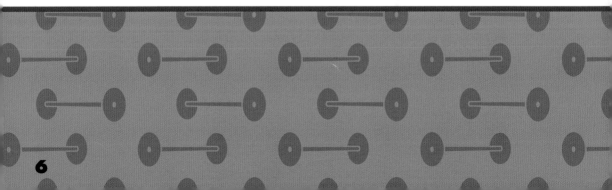

load

wheel

axle

Parts

A wheel is round.

An axle is a bar.

It connects to the wheel.

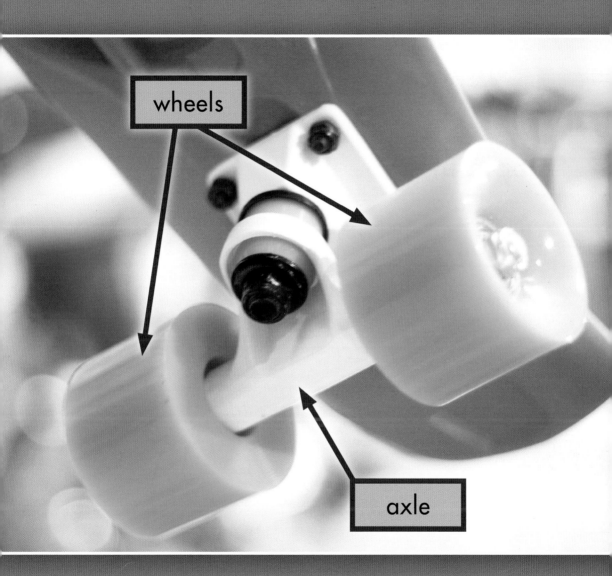

wheels

axle

Spin the wheel.

The bar turns.

Spin the bar.

The wheel turns.

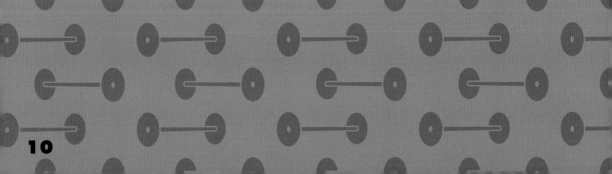

wheel

bar

The wheel and axle

work together.

Heavy loads can roll.

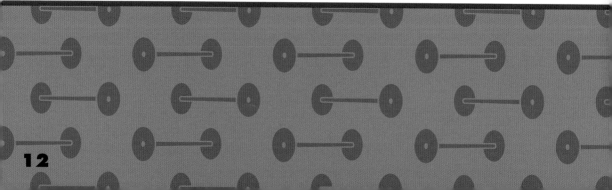

load

Everyday Tools

A Ferris wheel has

a wheel and axle.

Spin!

We ride around.

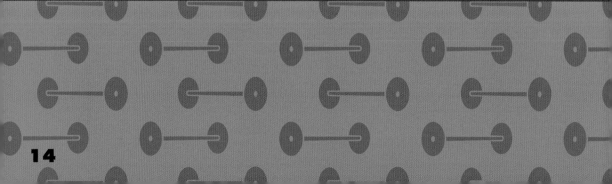

A knob has a wheel and axle.

Turn!

The door opens.

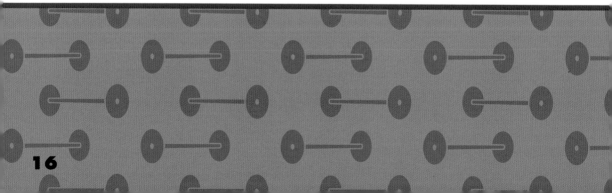

A wagon uses wheels
and axles.

Tug!

I pull my brother.

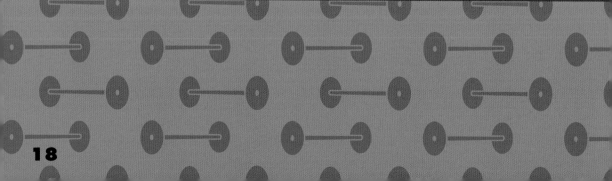

We use a simple machine.

It makes work easier and fun.

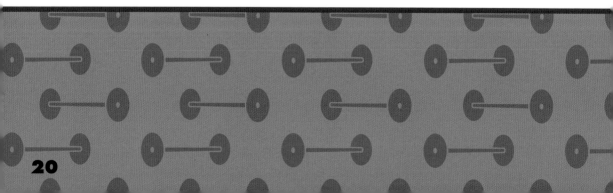

Glossary

axle—a rod or bar

knob—a round handle that opens a door

load—an object that you want to move or lift

simple machine—a tool that makes it easier to do something

tool—an item used to make work easier

wheel and axle—a simple machine with a circle connected to a rod

work—a job that must be done

Read More

Rivera, Andrea. *Wheels and Axles.* Simple Machines. Minneapolis: Abdo Zoom, 2017.

Schuh, Mari. *Hauling a Pumpkin: Wheels and Axles vs. Lever.* Simple Machines to the Rescue. Minneapolis: Lerner Publications, 2016.

Weakland, Mark. *Fred Flintstone's Adventures with Wheels and Axles: Bedrock and Roll!* Flintstones Explain Simple Machines. North Mankato, Minn.: Capstone Press, 2016.

Internet Sites

Use FactHound to find Internet sites related to this book.

Visit www.facthound.com

Just type in 9781543500769 and go.

Super-cool stuff!

Check out projects, games and lots more at
www.capstonekids.com

Critical Thinking Questions

1. What does the axle connect to?

2. What happens to the wheel if you spin the bar?

3. What types of wheels and axles do you like to use? How do they help you?

Index